WILLIAM WOODVILLE, M.D. F.R.S.

Author of Medical Botany, &c.

View of the inoculating Hospital at Pancras

Title: **MEDICAL & OFFICINAL PLANTS - VOL. 2 By William Woodville & James Sowerby XIX centuries engravings**
Series realized by Luca S. Cristini. Scientific consulence by Marco Rampinelli.

ISBN code: 978-88-93272292 First edition March 2017
Code.: **MUSEUM-005**, Editorial series code: Darwin's View **DV-004**

Cover & Art Design: Luca S. Cristini & Anna Cristini
MUSEUM is a trademark of Soldiershop publishing, via Padre Davide, 7 - 24050 Zanica (BG) ITALY. www.bookmuseum.it

WILLIAM WOODVILLE - JAMES SOWERBY

MEDICAL & OFFICINAL PLANTS - VOL. 2

PIANTE OFFICINALI, MEDICINALI E AROMATICHE

Shown in this series of three books are the complete and original pattern hand-colored engravings plates from the artist James Sowerby's medical plants present in the great work of William Woodville: Medical botany (London : Printed and sold for the author, by James Phillips, 1790-1793). Medical Botany, William Woodville's three volume work of materia medica, was published in monthly installments between 1790 and 1793.
A third edition of five volumes (the same used in our reproduction) was presented in 1832, twenty-seven years after Woodville's death. This publication added descriptions of thirty-nine new plants and was edited and revised by the eminent botanist, William Jackson Hooker (1785-1865).
With this work, Woodville intended to educate medical practitioners about the plants they prescribe and improve upon preceding works by introducing new plants and more detail.

William Woodville was an English physician and botanist who lived and worked in London for most of his life. For his work in botany, he was made a Fellow of the Linnaean Society just a year after the publication of volume one of his Medical Botany.

MEDICAL BOTANY

Containing systematic and general descriptions, with plates, of all the medicinal plants, indigenous and exotic, comprehended in the catalogues of the *materia medica*, as published by the Royal colleges of physicians of London and Edinburgh: accompanied with a circumstantial detail of their medicinal effects, and of the diseases in which they have been most successfully employed.

By William Woodville, M. D. Of the Royal college of physicians, London. Printed and Sold for the Author, by James Phillips, in five volumes (Edition of 1832)

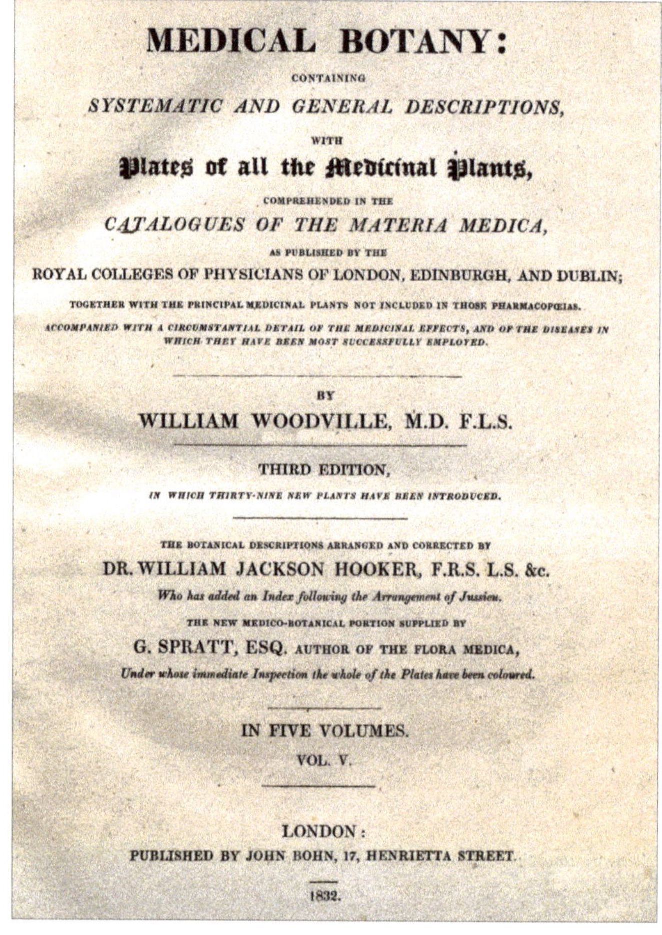

Medicus omnium Stirpium (si fieri potest) peritiam habeat; sin minus plurium saltem quibus frequenter utimur.   *Galen, Lib. De Antidot.*

PREFACE

In the catalogues of the *Materia Medica*, the productions of the animal and mineral kingdoms bear a small proportion to those of the vegetable.

Though it must be acknowledged that for some time past the medicinal uses of vegetable simples have been less regarded by physicians than they were formerly, which probably may be ascribed to the successive discoveries and improvements in chemistry;

it would however be difficult to shew that this preference is supported by any conclusive reasoning drawn from a comparative superiority of Chemicals over *Galenicals*, or that the more general use of the former has actually led to a more successful practice.

Although what may be called the herbaceous part of the *Materia Medica*, as now received in the British pharmacopoeias, comprises but a very inconsiderable portion of the vegetable world; yet limited as it now is, few medicinal practitioners have a distinct botanical knowledge of the individual plants of which it is composed, though generally, well acquainted with their effects and pharmaceutical uses.

But the practitioner, who is unable to distinguish those plants which he prescribes, is not only subjected to the impositions of the ignorant and fraudulent, but must feel a dissatisfaction which the inquisitive and philosophic mind will be anxious to remove, and to such it is presumed Medical Botany, by collecting and supplying the information necessary on this subject, will be found an acceptable and useful work; the professed design of which is not only to enable at the reader to distinguish with precision all those plants which are directed for medicinal use by the Colleges of London and Edinburgh, but to furnish him at the same time with a circumstantial detail of their respective virtues, and of the diseases in which they have been most successfully employed by different writers.

A distinctive and characteristic knowledge of natural objects mould certainly precede the consideration of their different properties and qualities; and with respect to plants, this knowledge is seldom to be adequately attained by a mere verbal description: accurate delineations therefore, become necessary, and this department is committed to Mr, Sowerby, an artist of established reputation, whole talents are not less conspicuous in the correctness than in the beauty of his designs.

It is justly a matter of surprise, that notwithstanding the universal adoption of the Linnaean system of Botany, and the great advances made in natural science, the works of Blackwell and Sheldrake should still be the only books in this country in which copper-plate figures of the medicinal plants are professedly given; while splendid foreign publications of them, by Regnault, Zorn, and Plenk, have appeared in the space of a very few years. These works however are far from superseding that now offered to the public; for without resorting to the invidious talk of pointing out their errors and imperfections, the author has the satisfaction of having exhibited Icons of several rare and valuable plants, which have never been completely figured in any preceding work whatever: and by subjoining some account of the botanical and medical history of each species, curiosity is more fully gratified, and a double interest is excited in the mind of the student.

Respecting the uses of Simples, the opinion of Oribasius will not be disputed, viz. " *Simplicium medicamentorum, & facultatum qua in eis insunt, cognitio ita necessaria est, ut sine ea nemo rite medicari queat* " and it is a lamentable truth, that our experimental knowledge of many of the herbaceous simples is extremely defective; for as writers on the *Materia Medica* have usually done little more than copy the accounts given by their predecessors, the virtues now ascribed to several plants are wholly referable to the authority of Dioscorides.

It is however hoped that the medical reader will find what relates to this part of the work as complete as the flow progressive state of experience in physic will admit: with this intention, facts and opinions have been industriously collected from various authorities; and those adduced by Professor Murray, and the works of the late Dr. Cullen, have furnished the largest contribution.

The publication of this work in monthly numbers has afforded the author an opportunity of knowing already the sentiments entertained of it, by several Gentlemen of great medical and botanical authority; from whose unsolicited communications he has derived considerable assistance, and for whose friendly suggestions he desires to make his most grateful acknowledgements.

(from the preface of first edition of Woodville work.)

Note on the plates

While today we may look at these pattern plate books as works of art, it is important to remember their initial function was to provide accurate representations of natural phenomena to promote the study of natural history. These notated hand-colored plates were meant to provide guidance in terms of color and shading to others involved in the printing process.

Details regarding the plates: copper plate engravings in black printing ink, hand colored with watercolor, with iron gall ink and graphite inscriptions.

Duplex est dos libelli.

BIOGRAPHIES

William Woodville (1752-1805) was a scientist born into a Quaker family at Cockermouth, Cumberland. He stusies medicine at the university of Edinburgh in Scotland, graduating MD on 12 September 1775. He start his work in European mainland before entering practice in his native country. After he became a famous physician and in 1791 was elected director of the prestigious Smallpox and Inoculation hospitals, St. Pancras in London. With this assignment he appropriated two acres at Battle bridge , near the hospital, where he established a botanical garden, which was maintained at his expenses. In the same year he realize the three volume treatise, *Medical Botany*. In this work , Woodville described, with illustrations and accounts of therapeutic effects all know medicinal plants at this time.

▲ William Woodville M.D. portrait

▼ James Sowerby portrait by Heaphy (1816)

James Sowerby (1757-1822) was an English naturalist and illustrator. Trained at the Royal Academy He gained a reputation as a fine illustrator for his contributions to works on botany and other fields of natural science. His reputation was such that he later was involved in several multi-volumed sets published over several years. His family, including both sons and daughters, were involved in his publication endeavors.

MEDICINAL PLANTS

Medicinal plants have been identified and used from prehistoric times. Plants make many chemical compounds for biological functions, including defence against insects, fungi and herbivorous mammals. Over 12,000 active compounds are known to science. These chemicals work on the human body in exactly the same way as pharmaceutical drugs, so herbal medicines can be beneficial and have harmful side effects just like conventional drugs. However, since a single plant may contain many substances, the effects of taking a plant medicine can be complex.

The earliest historical records of herbs are found from the Sumerian civilisation, where hundreds of medicinal plants including opium are listed on clay tablets. The Ebers Papyrus from ancient Egypt describes over 850 plant medicines. Drug research makes use of ethnobotany to search for pharmacologically active substances in nature, and has in this way discovered hundreds of useful compounds. These include the common drugs aspirin, digoxin, quinine, and opium. The compounds found in plants are of many kinds, but most are in four major biochemical classes, the alkaloids, glycosides, polyphenols, and terpenes.

HISTORY

Plants, including many now used as culinary herbs and spices, have been used as medicines from prehistoric times. Spices have been used partly to counter food spoilage bacteria, especially in hot climates, and especially in meat dishes which spoil more readily. Angiosperms (flowering plants) were the original source of most plant medicines. Human settlements are often surrounded by weeds useful as medicines, such as nettle, dandelion and chickweed. Some animals such as non-human primates, monarch butterflies and sheep ingest medicinal plants to treat illness.

Plant samples from prehistoric burial sites are among the lines of evidence that Paleolithic peoples had knowledge of herbal medicine. For instance, a 60.000-year-old Neanderthal burial site, "Shanidar IV", in northern Iraq has yielded large amounts of pollen from 8 plant species, 7 of which are used now as herbal remedies. The deliberate placement of flowers has been challenged. Paul B. Pettitt has stated that the *"deliberate placement of flowers has now been convincingly eliminated"*, noting that *"A recent examination of the microfauna from the strata into which the grave was cut suggests that the pollen was deposited by the burrowing rodent Meriones persicus, which is common in the Shanidar microfauna and whose burrowing activity can be observed today"*. A mushroom was found in the personal effects of *Ötzi the Iceman*, whose body was frozen in the Ötztal Alps for more than 5,000 years. The mushroom was probably used to treat whipworm.

ANCIENT TIMES

In ancient Sumeria, hundreds of medicinal plants including myrrh and opium are listed on clay tablets. The ancient Egyptian Ebers Papyrus lists over 800 plant medicines such as aloe, cannabis, castor bean, garlic, juniper, and mandrake.

From ancient times to the present, Ayurvedic medicine as documented in the Atharva Veda, the Rig Veda and the Sushruta Samhita has used hundreds of pharmacologically active herbs and spices such as turmeric, which contains curcumin. The Chinese pharmacopoeia, the *Shennong Ben Cao Jing* records plant medicines such as chaulmoogra for leprosy, ephedra, and hemp. This was expanded in the Tang Dynasty *Yaoxing Lun*.

In the fourth century BC, Aristotle's pupil Theophrastus wrote the first systematic botany text, *Historia plantarum*. In the first century AD, the Greek physician Pedanius Dioscorides documented over 1000 recipes for medicines using over 600 medicinal plants in *De materia medica*; it remained the authoritative reference on herbalism for over 1500 years, into the seventeenth century.

MIDDLE AGES

In the Early Middle Ages, Benedictine monasteries preserved medical knowledge in Europe, translating and copying classical texts and maintaining herb gardens. Hildegard of Bingen wrote *Causae et Curae* ("Causes and Cures") on medicine.

In the Islamic Golden Age, scholars translated many classical Greek texts including Dioscorides into Arabic.

Herbalism flourished in Baghdad and in Al-Andalus. Abulcasis (936–1013) of Cordoba wrote *The Book of Simples*, and Ibn al-Baitar (1197–1248) recorded hundreds of medicinal herbs such as *Aconitum*, nux vomica, and tamarind in his *Corpus of Simples*. Avicenna included many plants in his 1025 *The Canon of Medicine*. Abu-Rayhan Biruni, Ibn Zuhr, Peter of Spain, and John of St Amand wrote further pharmacopoeias.

EARLY MODERN

The early modern period saw the flourishing of illustrated herbals across Europe, starting with the 1526 *Grete Herball*. John Gerard wrote his famous *The Herball or General History of Plants* in 1597, based on Rembert Dodoens, and Nicholas Culpeper published his *The English Physician Enlarged*. Many new plant medicines arrived in Europe as products of Early Modern exploration and the resulting Columbian Exchange. In Mexico, the sixteenth century *Badianus Manuscript* described medicinal plants available in Central America.

PIANTE OFFICINALI

Una **pianta officinale** è un organismo vegetale usato nelle officine farmaceutiche per la produzione di specialità medicinali. Sono considerate piante officinali piante medicinali, aromatiche e da profumo inserite negli elenchi specifici e nelle farmacopee dei singoli paesi. Il numero e il tipo di piante officinali varia da paese a paese a seconda delle tradizioni. Il più comune utilizzo di piante officinali è quello di correttori del gusto: molti farmaci o preparati farmaceutici hanno originariamente un gusto sgradevole, che quindi viene "corretto" con l'aggiunta di sostanze di origine vegetale. Le piante officinali, ad esempio, sono quelle usate per conferire a sciroppi o a caramelle il gusto di fragola, arancia, limone, ecc.

Nel linguaggio comune spesso si sovrappone l'uso dei termini pianta medicinale con pianta officinale, termini che legalmente indicano due diverse entità; il termine officinale è un termine esclusivamente procedurale e indica quelle piante inserite all'interno di elenchi ufficiali come utilizzabili dalle officine farmaceutiche, a prescindere dal fatto che queste piante abbiano o meno proprietà di tipo medicinale. Il termine pianta medicinale indica invece quelle piante che contengono sostanze utilizzabili direttamente a scopo terapeutico o come precursori in emisintesi che portino a sostanze attive. È quindi chiaro che una pianta può essere officinale in un paese e non in un altro, a seconda delle regolamentazioni, ma essa sarà una pianta medicinale a prescindere dalle leggi. Una **pianta medicinale**, secondo l'Organizzazione Mondiale della Sanità (OMS), è un organismo vegetale che contiene in uno dei suoi organi sostanze che possono essere utilizzate a fini terapeutici o che sono i precursori di emisintesi di specie farmaceutiche.

Si va sempre più affermando il concetto di *fitocomplesso*, quale insieme di sostanze di origine vegetale non riproducibili per sintesi chimica. Il fitocomplesso va inteso come l'insieme di una quantità di principi attivi, noti e non, farmacologicamente attivi, e di sostanze che aiutano l'azione dei primi, pur essendo di per sé queste ultime farmacologicamente inattive. L'insieme delle interazioni dei primi (i principi attivi) e dei secondi (i coadiuvanti) determina le azioni note del fitocomplesso. Sempre secondo l'OMS circa il 25% dei moderni farmaci usati in USA sono di origine vegetale;inoltre sono 7.000 circa i composti medici, presenti nella moderna farmacopea, derivati da piante.

PIANTA MEDICINALE E OFFICINALE

Nel linguaggio comune si sovrappone l'uso dei termini pianta medicinale con pianta officinale, che indica piante utilizzate nelle officine farmaceutiche per la produzione di specialità medicinali. Questa definizione è però abbastanza riduttiva, e l'utilizzo in ambienti accademici del termine pianta medicinale non fa più riferimento esclusivamente ad un utilizzo a scopo terapeutico delle sostanze contenute nelle piante, bensì dell'utilizzo della pianta o di estratti da essa derivati a scopo terapeutico.

PIANTE IN PERICOLO

Un'indagine scientifica internazionale promossa dall'OMS all'inizio degli anni novanta, ha rilevato un numero di circa sessantamila specie vegetali, utilizzabili per la cura delle malattie, in forte pericolo di estinzione, di cui trecentosettantaquattro in Italia. Questo fatto richiede una maggiore attenzione alle piante medicinali, non solo quelle utilizzate nelle emisintesi, ma anche quelle che forniscono naturalmente componenti attivi applicabili nell'ambito della fitoterapia.

CENNI STORICI

Con l'introduzione dell'agricoltura si rese necessaria una maggiore attenzione alla vita delle piante e questo fu il punto di partenza della conoscenza, anche medica, delle caratteristiche delle piante stesse. Il più antico documento medico, per ora rintracciato, è il *"papiro di Ebers"*, risalente al 1500 a.C. Gli egizi facevano largo uso di medicamenti di natura vegetale, in particolar modo conoscevano le proprietà della maggiorana, dell'edera, della mirra.

Nell'antica Grecia, le conoscenze sulle piante si mescolarono con le teorie filosofiche sulle stesse. Uno dei più importanti studiosi fu Eracleide, il quale sperimentò nuove ricette, riprese in seguito da Celso. Le radici studiate e messe in vendita vennero definite *"farmacopoli"* e si basavano soprattutto sulle nozioni contenute nei testi medici scritti da Ippocrate (V secolo a.C.) e in quelli botanici scritti da Teofrasto.

Nell'antica Roma, già nel I secolo d.C. vennero impiantati orti chiamati medicinali, in quanto si coltivavano piante sfruttate per le varie terapie mediche.

Nel IX secolo d.C., in Sicilia, grazie ai Saraceni furono introdotte nuove tecniche idrauliche e di irrigazione che consentirono l'introduzione di nuove piante officinali. Gli arabi diedero un grande impulso sia all'alchimia sia alla chimica, che ebbe ripercussioni nello sviluppo farmaceutico di tinture e distillati. Gli arabi furono i primi ad organizzare una farmacopea, quindi un elenco di ricette descriventi le proporzioni e le composizioni chimiche. Ai secoli XI, XII, XIII, risalgono i primi testi farmaceutici, in cui confluirono le influenze greche, romane e arabe, sintetizzate nella definizione delle operazioni fondamentali: lozione, decozione, infusione e triturazione. In questo periodo si diffuse l'uso delle spezie e delle droghe e la Scuola salernitana introdusse assieme alle pratiche chirurgiche anche un antesignano dell'anestesia, la *spongia sonnifera*, imbevuta di oppio, succo di mandragora e di giusquiamo che doveva essere aspirata dal paziente. La Scuola di Salerno si distinse anche per la grande perizia nel selezionare le erbe, sulle quali abbondano indicazioni terapeutiche che si sono dimostrate efficaci ancora ai nostri tempi, valga per tutte l'insegnamento che diceva: «*Purga l'isopo dalle flemme il petto*», che ha un'azione benefica sulle bronchiti e sulle affezioni respiratorie.

La botanica intesa come scienza nacque solo agli inizi del Cinquecento, grazie alle scoperte geografiche e alla introduzione della stampa. Si diffusero, in questo periodo i primi erbari secchi e nel 1533 a Padova fu istituita la prima cattedra di "botanica sperimentale".

Pietro Andrea Mattioli redasse nel 1554 il più significativi testo di medicina e di botanica dell'epoca. Nel Seicento Pierre Magnol inserì nella classificazione l'intuizione delle famiglie, suddividendo il mondo vegetale in settantasei gruppi.

Nel secolo successivo una grande spinta al progresso della botanica fu effettuata dallo svedese Carl von Linné, che identificò le specie viventi dividendole in basi alle classi, agli ordini e ai generi.

Da allora l'evoluzione è stata continua.

2
THE PLATES
LE TAVOLE

PLANTS, MEDICINAL

HAND-COLORED ENGRAVING BY

JAMES SOWERBY

PUBLISHED BY

DR. WOODVILLE.

PLATES LIST OF ILLUSTRATIONS

154. Trigonella Foenum graecum - Penugreek
155. Astragalus Exscapus - Steemless milk vetch
156. Pierocarpus Santalinus - Red Saunders Tree
157. Mimosa Catechu- Catechu Mimosa
158. Mimosa Nilotica- Egyptian Mimosa, Acacia, Egyptian Thorn
159. Cassia Senna - Senna or EgyptianCassia
160. Cassia Fistula- Purging Cassia
161. Tamarindus Indica - Tamarind Tree
162. Polygala Senega - Rattle-snake root milk wort
163. Haematoxylum Campechianum- Logwood
164. Fumaria Officinalis - Common Fumitory
165. Aconitum Lapellus- Common wolf's bane or Monk's hood
166. Dictamnus Albus - White Fraxinella or bastard Dittany
167. Anemone Pratensis - Meadow Anemone or Pasque Flower
168. Delphinium Staphisagria - Palmated Larkspur or Stavesacre
169. Helleborus Niger - Black Hellebore or Christmas Rose
170. Helleborus Faetidus - Fetid Hellebor or bear's foot
171. Clematis Recta - Upright Virgin's bower
172. Ranuculus Acris - Upright Meadow Crowfoot
173. Paeonia Officilanis - Common Peony
174. Ruta Graveolens - Common Rue
175. Potentilla Reptans – Common Cinquefoil
176. Rubus Idaeus - Raspberry Bush
177. Rosa Canina - Dog Rose or Hep Tree
178. Rosa Centifolia - Hundred-leaved Rose
179. Rosa Gallica - Red Officinal Rose
180. Agrimonia Eupatoria – Common Agrimony
181. Geum Urbanus – Common Avens
182. Pyrus Cydonia - Common Quince Tree
183. Amygdalus Communis - Almond Tree
184. Amygdalus Persica - Common Peach Tree
185. Prunus Laurocerasus - Common or Cherry Laurel
186. Prunus Spinosa - Sloe Tree
187. Prunus Domestica - Common Prune or Plum Tree
188. Citrus Aurantium - Orange Tree
189. Citrus Medica - Lemon Tree
190. Punica Granatum - Pomegranate Tree
191. Ribes Rubrum - Red Currant
192. Ribes Nigrum - Black Currant
193. Caryophyllus Aromaticus - Clove Tree
194. Myrtus Pimenta - Pimento, Jamaica Pepper, All-spice
195. Lelaleuca Leucadendron - Cajeput Tree or Aromatic Melaleuca
196. Sedum Acre – Wall stone-crop or wall Pepper
197. Saxifraga Granulata – White Saxifrage
198. Althaea Officinalis - Marsh-mallow
199. Malva Sylvestris - Common Mallow
200. Guaiacum Officinale - Officinal Guaiacum
201. Oxalis Acetosella – Wood-sorrel
202. Linum Usitatissimum – Common Flax
203. Quassia Simaruba- Simaruba Quassia
204. Quassia Amara - Bitter Quassia
205. Dianthus Caryophyllus- Clove Pink
206. Saponaria Officinalis - Soapwort
207. Cistus Creticus - Cretan Cistus
208. Hypericum Perforatum - Perforated St. John's wort
209. Fraxinus Ornus - Flowering Ash
210. Rhamnus Catharticus - Purging Buckthorn
211. Sambucus Nigra - Common black Elder

99-100-101-102- Bicornes Various
Order Bicornes

103- *Rhododendron Chrisamanthum- Yellow flowered Rhododendron*
Order Bicornes

104- *Pulmonaria Officinalis - Lungwort*
Order Asperifoliae

105-106-107-108 *Officinalis Various*
Order Asperifoliae

109- *Cynoglossum Officinalis - Common Hound's tongue*
Order Asperifoliae

110- *Borago Officinalis- Common Borage*
Order Asperifoliae

111- Glecoma Hederacea- Ground-ivy or Gill
Order Verticillatae

113-114-115-116 Officinalis Various
Order Verticillatae

117- Rosmarinus Officinalis - Common Rosemary
Order Verticillatae

Marrubium vulgare

Mentha piperita

Mentha viridis

118-120-121-122- Minth Various
Order Verticillatae

119- *Melissa Officinalis - Common Balm*
Order *Verticillatae*

123- *Origanum Vulgare - Wild Marjoram*
Order *Verticillatae*

124- Origanum Majorana - Sweet Marjoram
Order Verticillatae

Thymus vulgaris.

Salvia officinalis

Betonica officinalis

Teucrium Chamædrys

125- 127- 128-130 *Officinalis Various*
Order Verticillatae

131- *Gratiola Officinalis - Hedge Hyssop*
Order Personatae

132- *Veronica Beccabunga - Brooklime Speedwell*
Order Personatae

133- *Verbena Officinalis* - Common Vervain
Order Personatae

Thymus Serpyllum.

Origanum Dictamnus.

Veronica officinalis

Euphrasia officinalis

126-129-134-135- *Offinical Herbs Various*
Order Verticillatae and Personatae

136- Antirrhinum Linaria - Common Toad-flax
Order Personatae

137- *Vitex Agnus Castus - Chaste-tree*
Order Personatae

138- *Papaver Somniferum - White Poppy*
Order Rhoeades

139- Papaver Rhoeas - Red or corn Poppy
Order Rhoeades

Chelidonium majus.

Capparis spinosa.

Cochlearia officinalis.

Cardamine pratensis.

140- 141- 142- 143 *Offinical Herbs Various*
Order Rhoeades

144- *Sisymbrium Nasturtium - Water-cresses*
Order Siliquosae

145-146-147-148- *Offinical Herbs Various*
Order Siliquosae

149- *Astragalus Tragacantha - Goat's thorn milk vetch*
Order Papilionaceae

150- *Spartium Scoparium - Common Broom*
Order Papilionaceae

151- *Geoffroya Inermis - Smooth Geoffroya or bastard cabbage-tree*
Order Papilionaceae

152- *Glycyrrhiza Glabra - Common Liquorice*
Order *Papilionaceae*

153- *Dolichos Pruriens - Cowhage Dolichos*
Order *Papilionaceae*

154- Trigonella Foenum graecum - Penugreek
Order Papilionaceae

155- *Astragalus Exscapus - Steemless milk vetch*
Order *Papilionaceae*

156- Pterocarpus Santalinus - Red Saunders Tree
Order Papilionaceae

157- Mimosa Catechu- Catechu Mimosa
Order Lomentaceae

158- *Mimosa Nilotica- Egyptian Mimosa, Acacia, Egyptian Thorn*
Order Lomentaceae

159- *Cassia Senna - Senna or EgyptianCassia*
Order Lomentaceae

160- Cassia Fistula- Purging Cassia
Order Lomentaceae

161- *Tamarindus Indica* - *Tamarind Tree*
Order Lomentaceae

162 - *Polygala Senega - Rattle-snake root milk wort*
Order Lomentaceae

163- *Haematoxylum Campechianum- Logwood*
Order Lomentaceae

164 - *Fumaria Officinalis* - Common Fumitory
Order Lomentaceae

165 - *Aconitum Lapellus* - Common wolf's bane or Monk's hood
Order Multisiliquae

166 - *Dictamnus Albus - White Fraxinella or bastard Dittany*
Order Multisiliquae

167 - Anemone Pratensis - Meadow Anemone or Pasque Flower
Order Multisiliquae

168 - *Delphinium Staphisagria* - Palmated Larkspur or Stavesacre
Order Multisiliquae

169- *Helleborus Niger - Black Hellebore or Christmas Rose*
Order Multisiliquae

170- Helleborus Faetidus - Fetid Hellebor or bear's foot
Order Multisiliquae

171- *Clematis Recta - Upright Virgin's bower*
Order Multisiliquae

172- *Ranunculus Acris - Upright Meadow Crowfoot*
Order Multisiliquae

173- *Paeonia Officilanis - Common Peony*
Order Multisiliquae

174- *Ruta Graveolens - Common Rue*
Order Multisiliquae

176- *Rubus Idaeus* - *Raspberry Bush*
Order Senticosae

177.

177- *Rosa Canina - Dog Rose or Hep Tree*
Order Senticosae

178- Rosa Centifolia - Hundred-leaved Rose
Order Senticosae

179- *Rosa Gallica - Red Officinal Rose*
Order Senticosae

182- *Pyrus Cydonia - Common Quince Tree*
Order Pomaceae

183- *Amygdalus Communis - Almond Tree*
Order Pomaceae

184 - Amygdalus Persica - Common Peach Tree

Order Pomaceae

185- *Prunus Laurocerasus - Common or Cherry Laurel*
Order Pomaceae

186 - *Prunus Spinosa* - *Sloe Tree*
Order Pomaceae

187- *Prunus Domestica - Common Prune or Plum Tree*
Order Pomaceae

188 - Citrus Aurantium - Orange Tree
Order Pomaceae

189- *Citrus Medica - Lemon Tree*
Order Pomaceae

190- *Punica Granatum* - *Pomegranate Tree*
Order Pomaceae

191- *Ribes Rubrum - Red Currant*
Order *Pomaceae*

192- *Ribes Nigrum - Black Currant*
Order Pomaceae

193- *Caryophyllus Aromaticus - Clove Tree*
Order *Hesperideae*

194- *Myrtus Pimenta - Pimento, Jamaica Pepper, All-spice*
Order Hesperideae

195- *Lelaleuca Leucadendron - Cajeput Tree or Aromatic Melaleuca*
Order Hesperideae

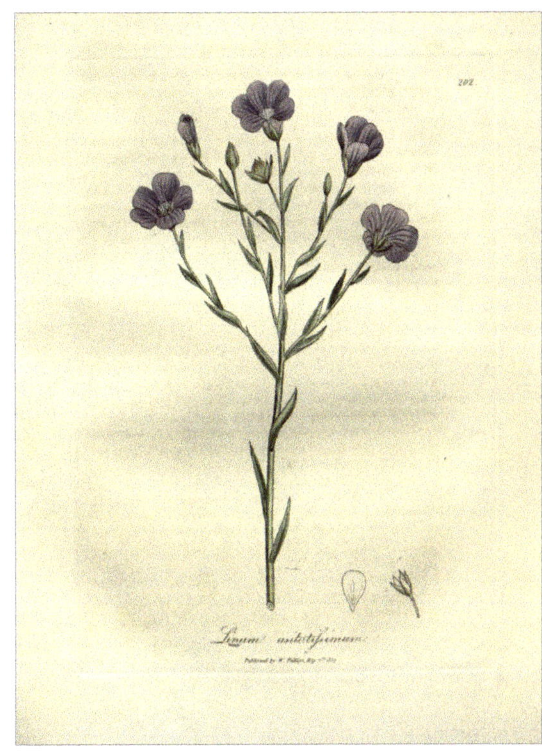

196-197-201-202- Succulentae and Gruinales Various
Order Succulentae and Gruinales

198- *Althaea Officinalis* - *Marsh-mallow*
Order *Columniferae*

199- *Malva Sylvestris - Common Mallow*
Order Columniferae

200- *Guaiacum Officinale - Officinal Guaiacum*
Order Gruinales

203- *Quassia Simaruba- Simaruba Quassia*
Order Gruinales

204 - *Quassia Amara* - *Bitter Quassia*
Order Gruinales

205- *Dianthus Caryophyllus- Clove Pink*
Order *Caryophylleae*

206- *Saponaria Officinalis - Soapwort*
Order Caryophylleae

207- Cistus Creticus - Cretan Cistus
Order Ascyroideae

208 - *Hypericum Perforatum* - Perforated St. John's wort
Order Ascyroideae

209- *Fraxinus Ornus - Flowering Ash*
Order Ascyroideae

210- *Rhamnus Catharticus - Purging Buckthorn*
Order Dumosae

211- *Sambucus Nigra - Common black Elder*
Order Dumosae

DARWIN'S VIEW SERIES

Actually, the world from Darwin's point of view. The new series of specifically dedicated to the animal, vegetable and mineral world. A great review of nature through his most beautiful and fascinating images, taken from ancient tomes and essays about nature, made   by the greatest individuals, artists and scientists together. Not only that, "Darwin's view" will involve yourself through the description of the stories, with facts and images of the exotic and romantic travels, made   by the great explorers and brilliant scientists of the past, starting with the epic one on the HMS Beagle of our beloved and legendary Charles Robert Darwin!

CONTENTS

www.ingramcontent.com/pod-product-compliance
Lightning Source LLC
LaVergne TN
LVHW071525180726
843512LV00014B/1168